At Sylvan, we believe that a lifelong love of learning begins at an early age, and we are glad you have chosen our resources to help your children experience the joy of mathematics as they build critical reasoning skills. We know that the time you spend with your children reinforcing the lessons learned in school will contribute to their love of learning.

Success in math requires more than just memorizing basic facts and algorithms; it also requires children to make sense of size, shape, and numbers as they appear in the world. Children who can connect their understanding of math to the world around them will be ready for the challenges of mathematics as they advance to more complex topics.

We use a research-based, step-by-step process in teaching math at Sylvan that includes thought-provoking math problems and activities. As students increase their success as problem solvers, they become more confident. With increasing confidence, students build even more success. The design of the Sylvan workbooks will help you to help your children build the skills and confidence that will contribute to success in school.

Included with your purchase of this workbook is a coupon for a discount at a participating Sylvan center. We hope you will use this coupon to further your children's academic journeys. Let us partner with you to support the development of confident, well-prepared, independent learners.

The Sylvan Team

Sylvan Learning Center
Unleash your child's potential here

No matter how big or small the academic challenge, every child has the ability to learn. But sometimes children need help making it happen. Sylvan believes every child has the potential to do great things. And, we know better than anyone else how to tap into that academic potential so that a child's future really is full of possibilities. Sylvan Learning Center is the place where your child can build and master the learning skills needed to succeed and unlock the potential you know is there.

The proven, personalized approach of our in-center programs delivers unparalleled results that other supplemental education services simply can't match. Your child's achievements will be seen not only in test scores and report cards but outside the classroom as well. And when your child starts achieving his or her full potential, everyone will know it. You will see a new level of confidence come through in all of your child's activities and interactions.

How can Sylvan's personalized in-center approach help your child unleash the potential you know is there?

• Starting with our exclusive Sylvan Skills Assessment®, we pinpoint your child's exact academic needs.

• Then we develop a customized learning plan designed to meet your child's academic goals.

• Through our method of skill mastery, your child will not only learn and master every skill in a personalized plan, but he or she will be truly motivated and inspired to achieve.

To get started, included with this Sylvan product purchase is $10 off our exclusive Sylvan Skills Assessment®. Simply use this coupon and contact your local Sylvan Learning Center to set up your appointment.

To learn more about Sylvan and our innovative in-center programs, call 1-800-EDUCATE or visit www.SylvanLearning.com. *With over 900 locations in North America, there is a Sylvan Learning Center near you!*

Kindergarten
Shapes &
Geometry Success

Copyright © 2011 by Sylvan Learning, Inc.

Published in the United States by Random House, Inc., New York, and in Canada by Random House of Canada Limited, Toronto.

www.tutoring.sylvanlearning.com

Created by Smarterville Productions LLC
Producer & Editorial Direction: The Linguistic Edge
Producer: TJ Trochlil McGreevy
Writer: Amy Kraft
Cover and Interior Illustrations: Shawn Finley, Tim Goldman, and Duendes del Sur
Layout and Art Direction: SunDried Penguin

First Edition

ISBN: 978-0-307-47925-9
ISSN: 2156-5953

This book is available at special discounts for bulk purchases for sales promotions or premiums. For more information, write to Special Markets/Premium Sales, 1745 Broadway, MD 6-2, New York, New York 10019 or e-mail specialmarkets@randomhouse.com.

PRINTED IN CHINA

10 9 8 7 6 5 4 3 2 1

Contents

Practice the Shape

TRACE the circles.

This shape is a **circle**.

"C" Is for Circle

WRITE the letter "C" in each circle.

Hide and Seek

DRAW a circle around every circle in the picture.

Mystery Shapes

COLOR yellow each section that has a circle. COLOR blue each section that does **not** have a circle.

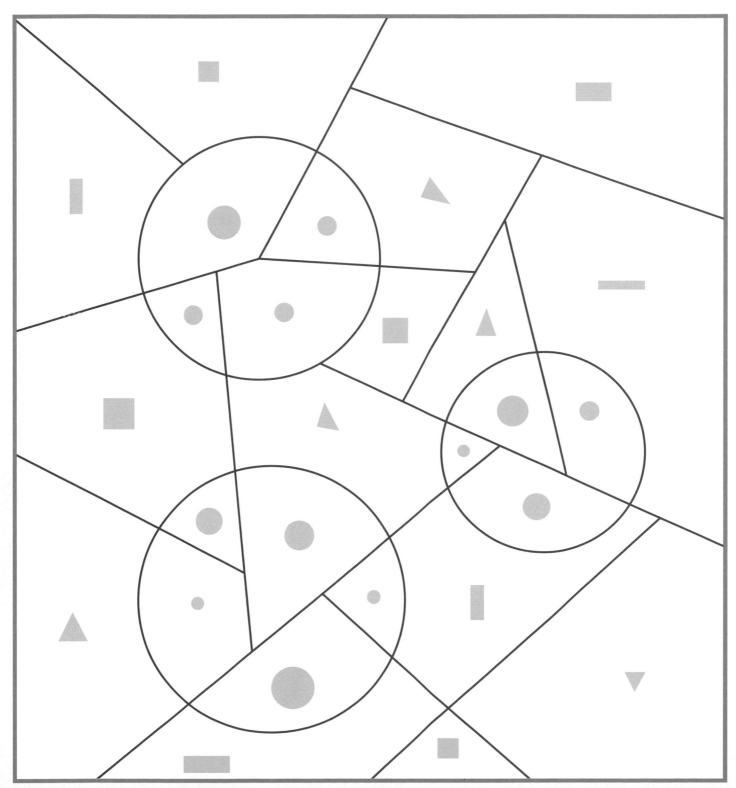

Cross Out

CROSS OUT any picture that is **not** a circle.

Doodle Pad

TRACE the circle. Then DRAW a picture using the circle.

Practice the Shape

TRACE the triangles.

This shape is a **triangle**.

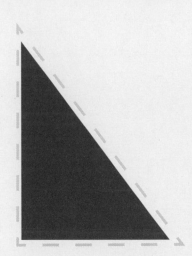

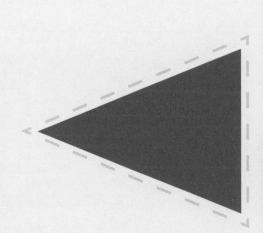

"T" Is for Triangle

WRITE the letter "T" in each triangle.

Hide and Seek

DRAW a triangle around every triangle in the picture.

Mystery Shapes

COLOR orange each section that has a triangle. COLOR purple each section that does **not** have a triangle.

Triangles

Cross Out

CROSS OUT any picture that is **not** a triangle.

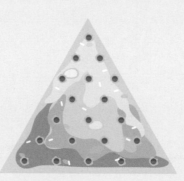

Doodle Pad

TRACE the triangle. Then DRAW a picture using the triangle.

Rectangles

Practice the Shape

TRACE the rectangles.

This shape is a **rectangle**.

"R" Is for Rectangle

WRITE the letter "R" in each rectangle.

Hide and Seek

DRAW a rectangle around every rectangle in the picture.

Mystery Shapes

COLOR green each section that has a rectangle. COLOR yellow each section that does **not** have a rectangle.

Cross Out

CROSS OUT any picture that is **not** a rectangle.

Doodle Pad

TRACE the rectangle. Then DRAW a picture using the rectangle.

Squares

Practice the Shape

TRACE the squares.

This shape is a **square**.

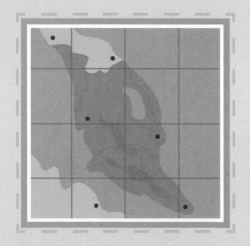

"S" Is for Square

WRITE the letter "S" in each square.

Hide and Seek

DRAW a square around every square in the picture.

Mystery Shapes

COLOR red each section that has a square. COLOR blue each section that does **not** have a square.

Cross Out

CROSS OUT any picture that is **not** a square.

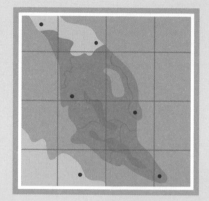

Doodle Pad

TRACE the square. Then DRAW a picture using the square.

Find the Same

CIRCLE the shape in each row that is the same shape as the first one.

Perfect Party

COLOR the party decorations according to the color of the shapes at the top.

Match Up

DRAW lines to connect the shapes that are the same.

Hide and Seek

How many of each shape can you find in the picture? WRITE the number next to each shape.

_____ 1

_____ 2

_____ 3

Spheres

Circle the Same

CIRCLE all of the pictures that have the same kind of shape as the top shape.

Hide and Seek

DRAW a circle around every object in the picture that matches the top shape.

Circle the Same

CIRCLE all of the pictures that have the same kind of shape as the top shape.

Cross Out

CROSS OUT any picture that does **not** match the top shape.

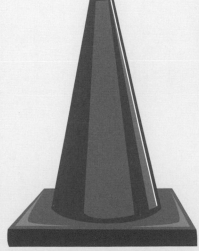

Rectangular Prisms

Circle the Same

CIRCLE all of the pictures that have the same kind of shape as the top shape.

Hide and Seek

DRAW a circle around every object in the picture that matches the top shape.

Circle the Same

CIRCLE all of the pictures that have the same kind of shape as the top shape.

Cross Out

CROSS OUT any picture that does **not** match the top shape.

Find the Same

CIRCLE the shape in each row that is the same shape as the first one.

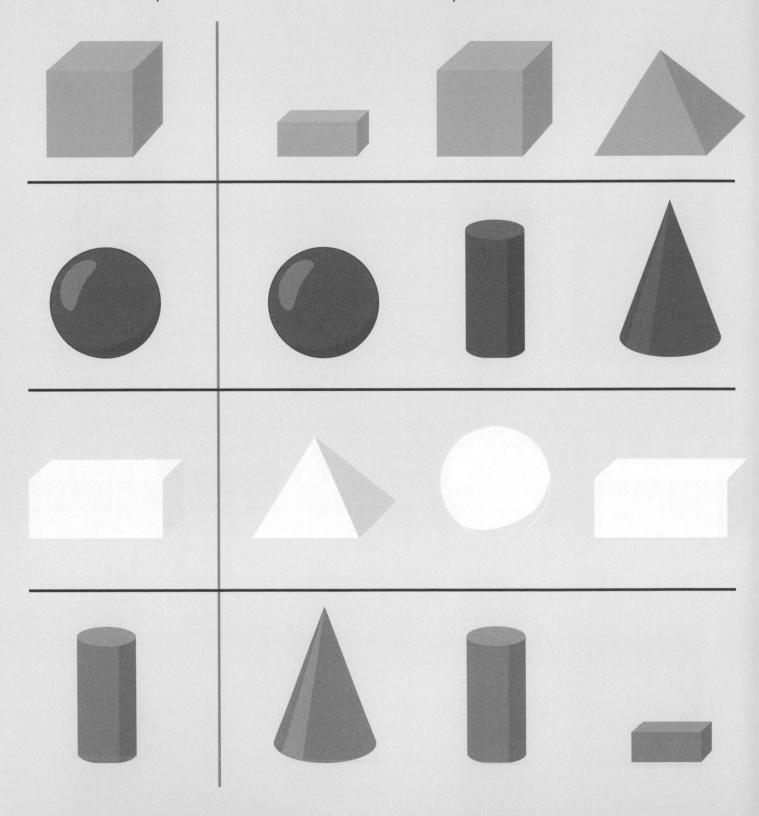

Nice Necklace

COLOR the beads in the necklace according to the color of the shapes at the top.

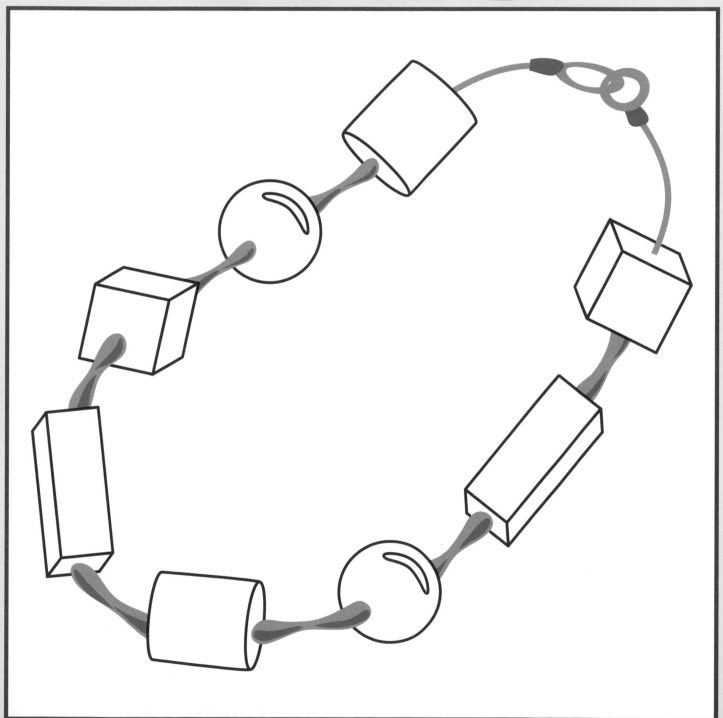

Match Up

DRAW lines to connect the shapes that are the same.

Hide and Seek

How many of each shape can you find in the picture? WRITE the number next to each shape.

1

2

3

Comparing Shapes

Odd One Out

CROSS OUT the picture in each row that does **not** go with the others.

   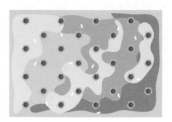

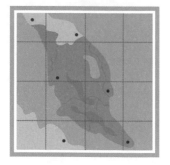   

42

Cheese and Crackers

CUT OUT the cards. READ the rules. PLAY the game!

Rules: Two players
1. Place the cards face down on a table.
2. Take turns turning over two cards at a time.
3. Keep the cards when you match a cracker with cheese that has the same shape.

The player with the most matches wins!

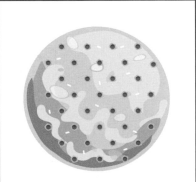

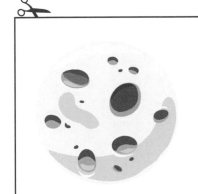
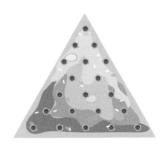

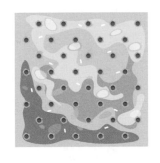

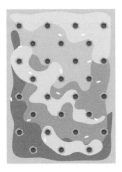

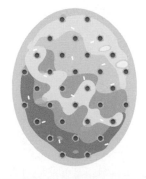

Find the Same

CIRCLE the shape in each row that is the same shape as the first one.

Comparing Shapes

Odd One Out

CROSS OUT the picture in each row that does **not** go with the others.

Find the Same

CIRCLE the shape in each row that is the same shape as the first one.

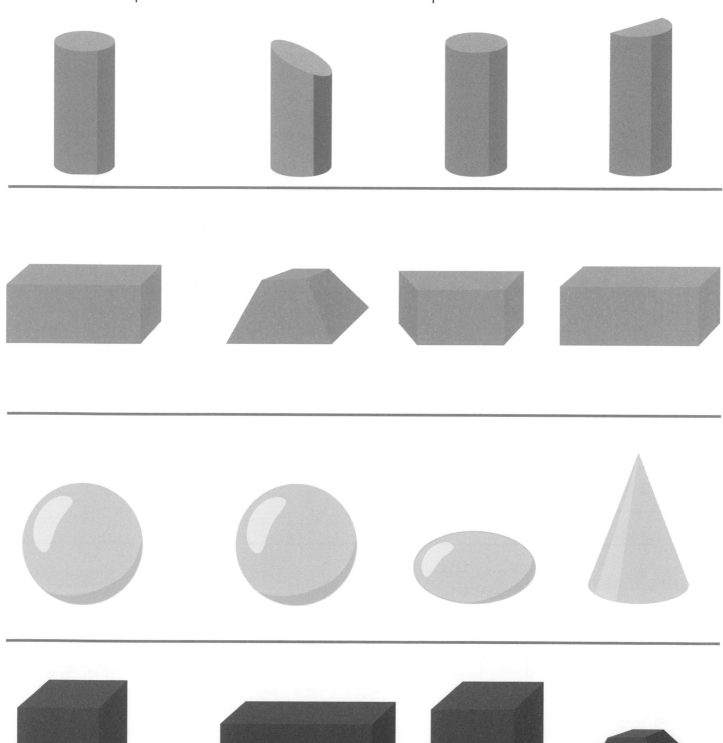

Who's Smaller?

COLOR the **smaller** shape in each pair.

Who's Bigger?

COLOR the **bigger** shape in each pair.

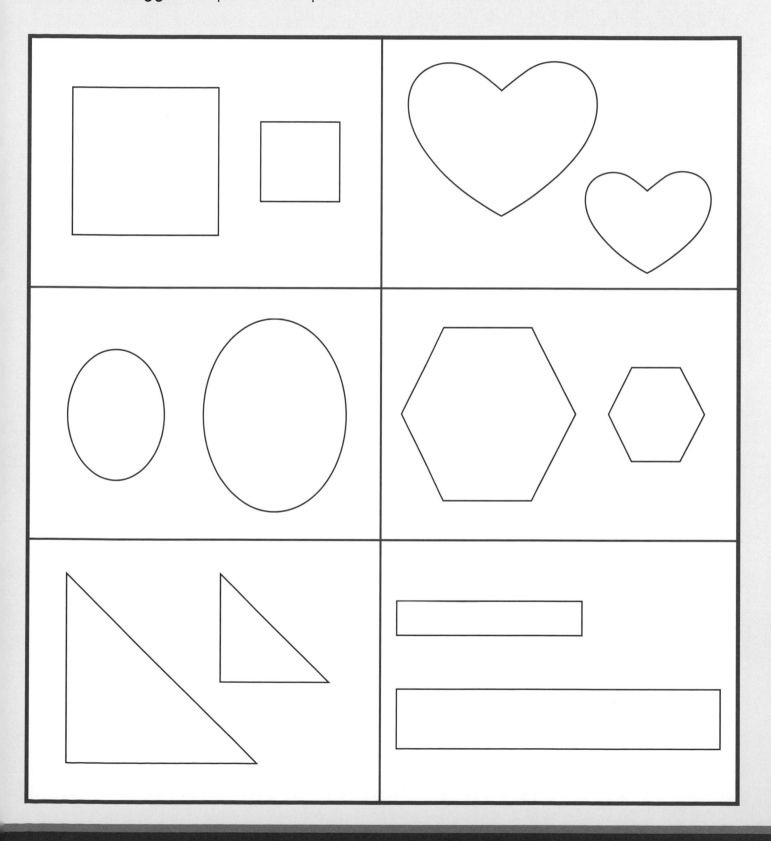

Inside and Out

DRAW a smaller version of each shape inside the shape. Then DRAW a bigger version of each shape outside the shape.

Soccer Size

CIRCLE each ball that is the same size as the soccer ball.

Biggest and Smallest

DRAW a line under the biggest picture in each row. Then DRAW a circle around the smallest picture in each row.

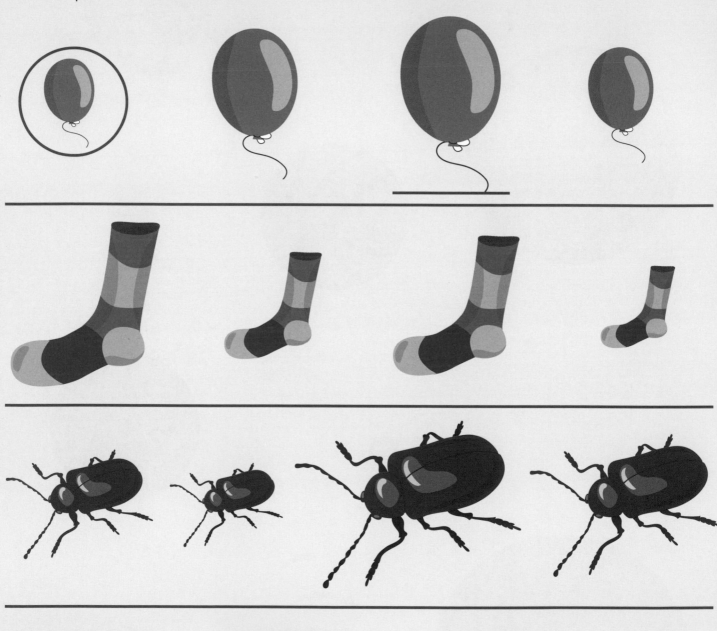

Picking Pairs

DRAW a line to connect each pair of objects that have the same shape.

Picking Pairs

DRAW a line to connect each pair of objects that have the same shape.

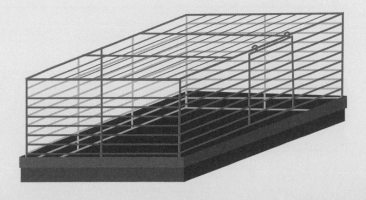

Small, Smaller, Smallest

CIRCLE the **smaller** shape.

CIRCLE the **smallest** shape.

Big, Bigger, Biggest

CIRCLE the **bigger** shape.

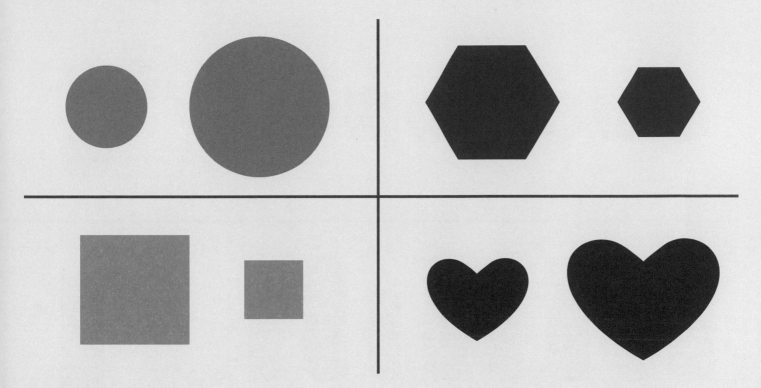

CIRCLE the **biggest** shape.

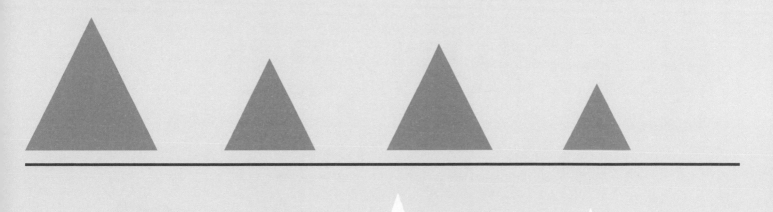

Shape Patterns

What Comes Next?

CIRCLE the shape that comes next in the pattern.

1.

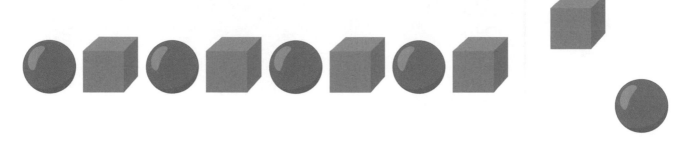

2.

3.

4.

5.

6.

7.

8,

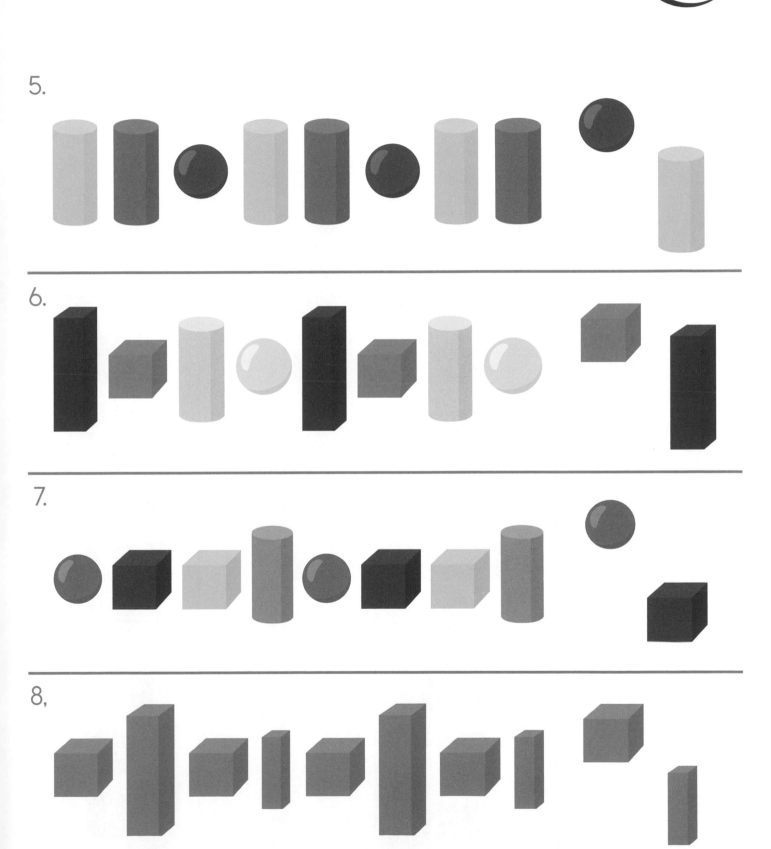

Shapely Sequence

DRAW and COLOR shapes to finish each pattern.

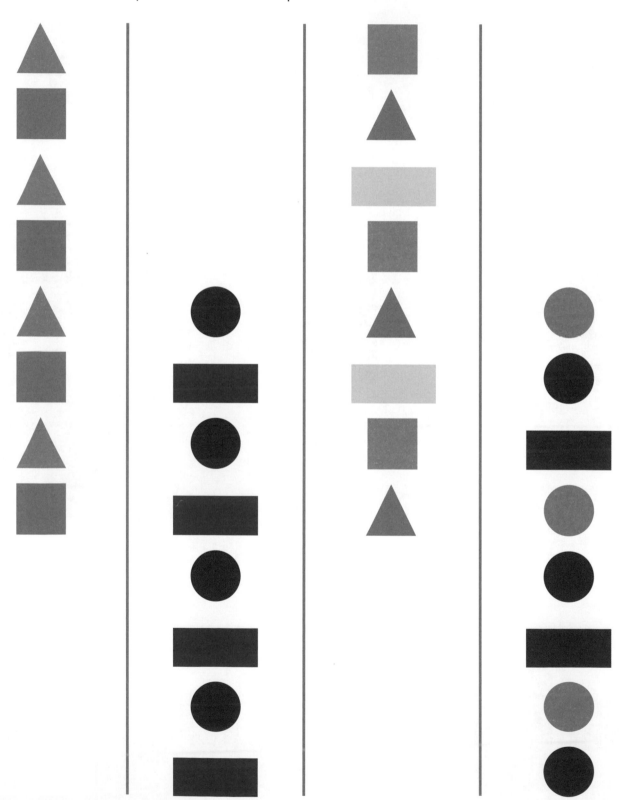

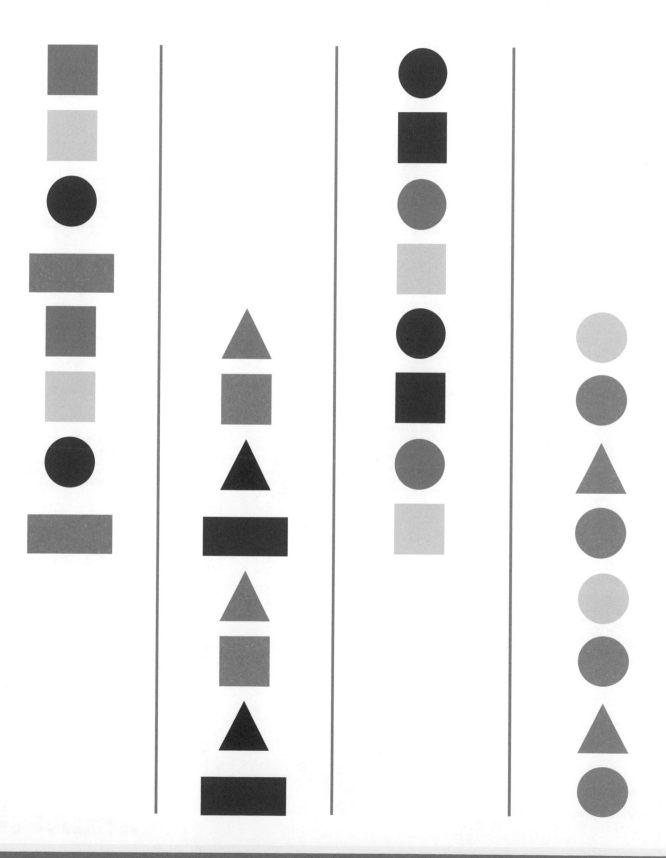

Missing Pieces

DRAW and COLOR the shapes to finish each pattern.

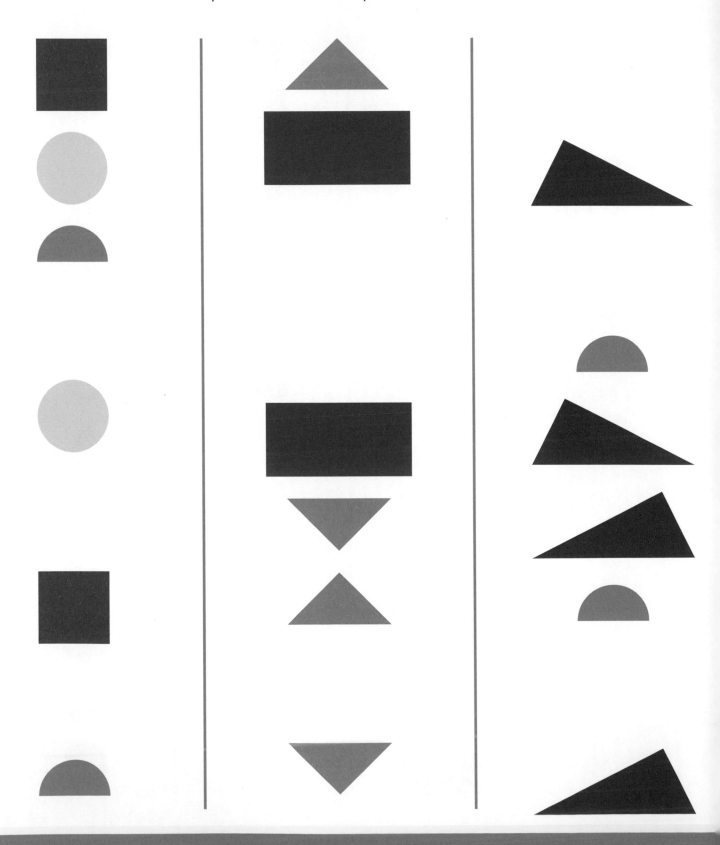

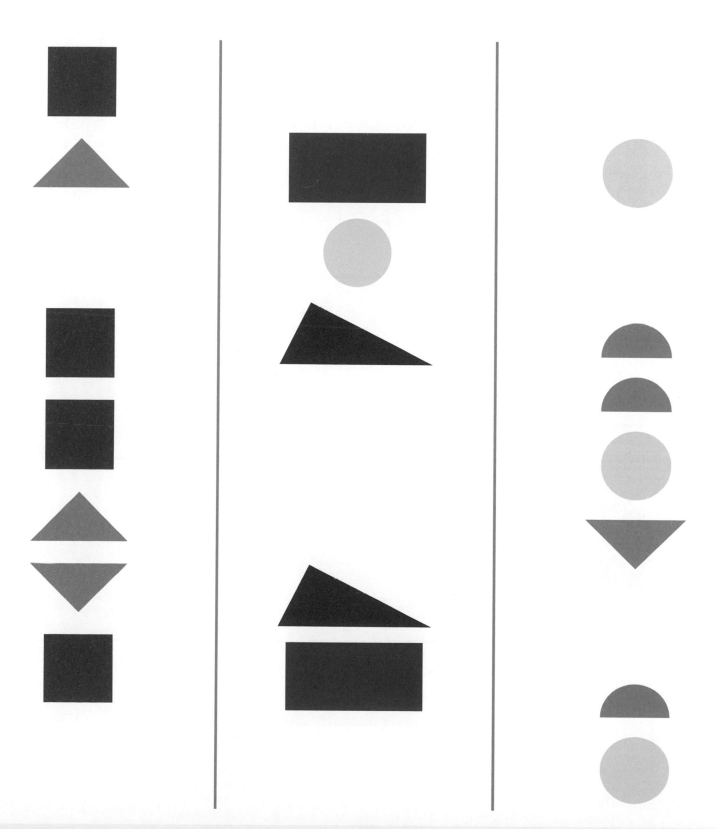

Matched Set

Using the yellow cards from page 65, PLACE two cards next to each shape to make a matched set. (Save the cards to use again.)

HINT: Look for things the pictures have in common, like shape, color, or anything else! See how many different sets you can make.

CUT OUT these 16 cards. The yellow cards are for use on page 64, and the blue cards are for use on page 67.

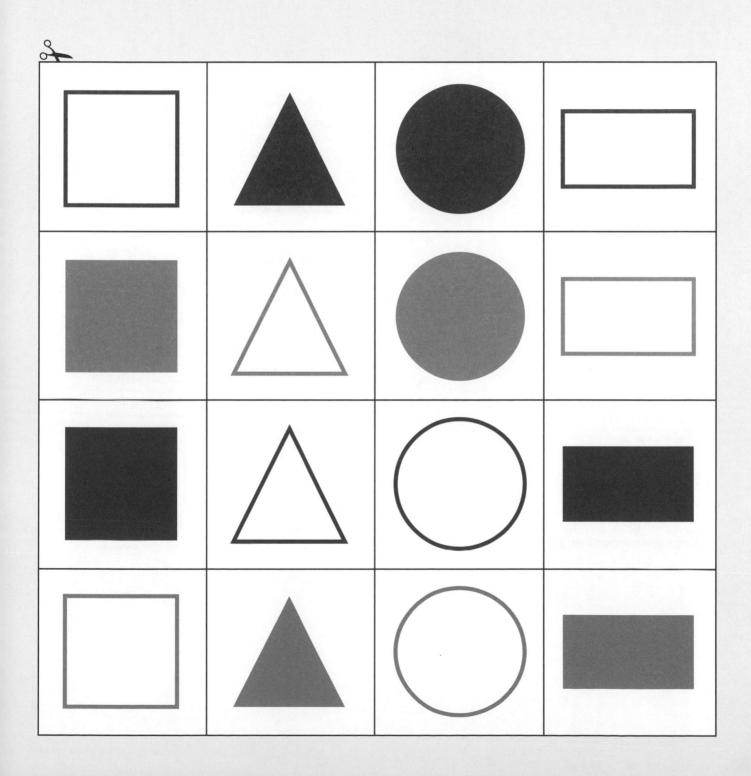

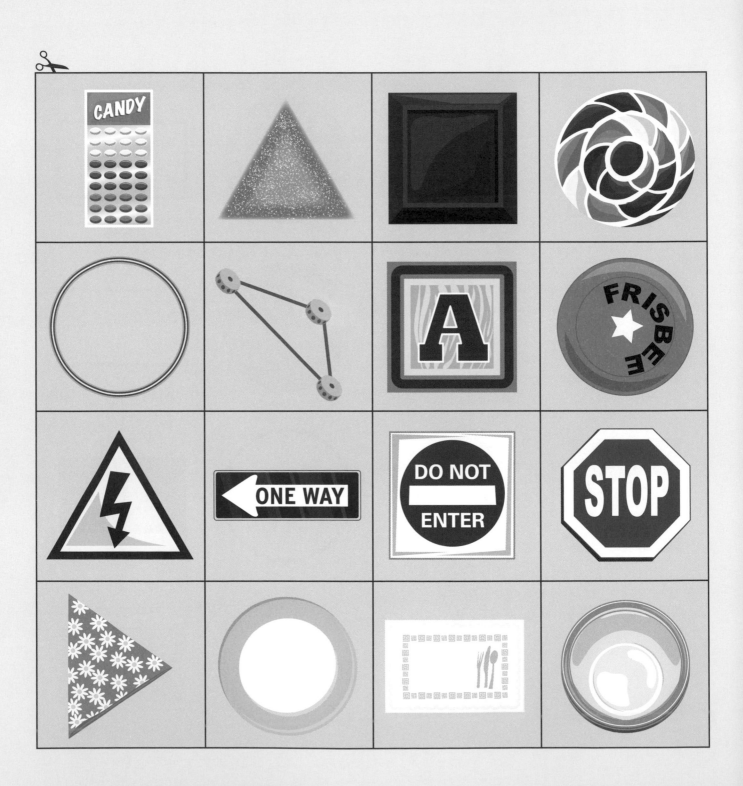

Matched Set

Using the blue cards from page 66, PLACE two cards next to each picture to make a matched set.

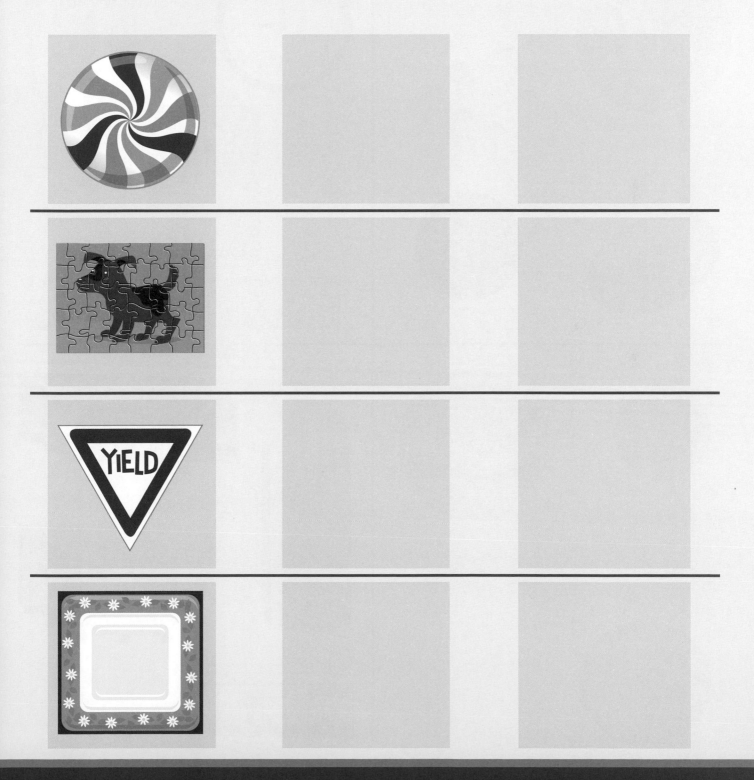

Shape Sets

Cross Out

CROSS OUT the picture in each set that does **not** go with the others.

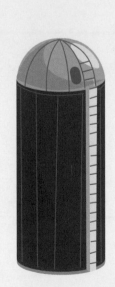

Finish the Set

DRAW and COLOR a shape to finish each set.

HINT: There is more than one correct answer for each set. For an extra challenge, see how many shapes you can add to each set.

Off the Wall

DRAW a line from each thing to the place where it should hang on the wall.

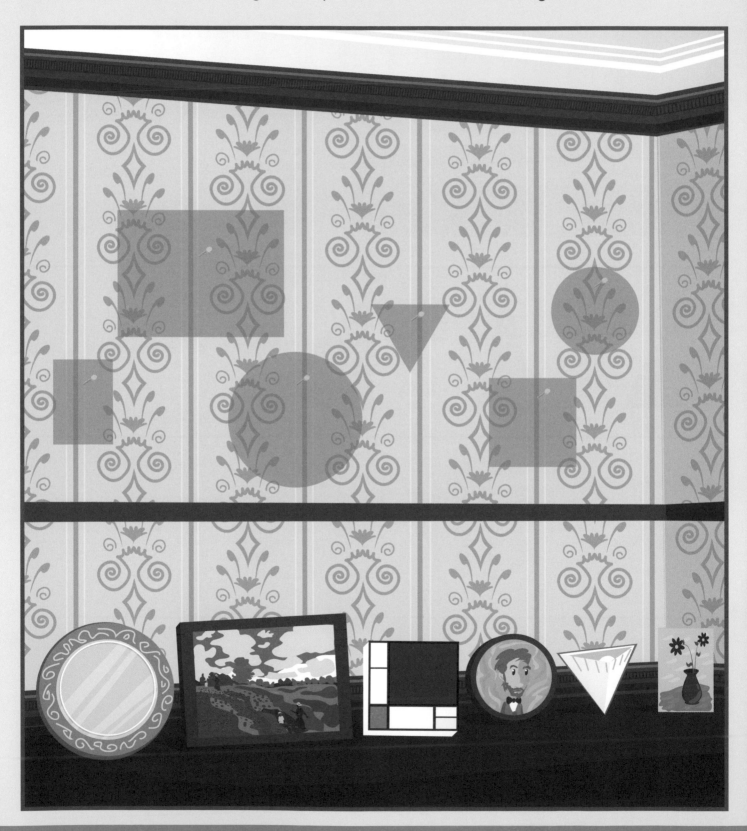

Playing Pieces

DRAW a line from the playing pieces to the game board where they belong.

HINT: Look at the shape of the game boards and the shape of the pieces.

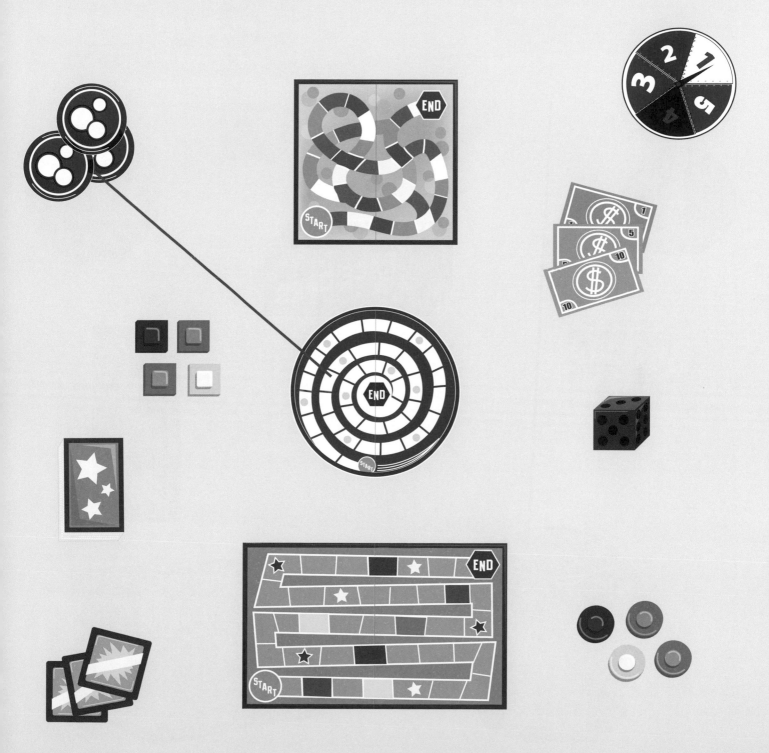

Put It Away

DRAW a line to put each toy in the toy box with the same shape.

Get in Place

DRAW a line from each shape outside the diagram to show where it belongs inside the diagram.

Sorting Shapes

Up the Flagpole

DRAW a line from each flag to the correct flagpole.

HINT: Look for similar shapes.

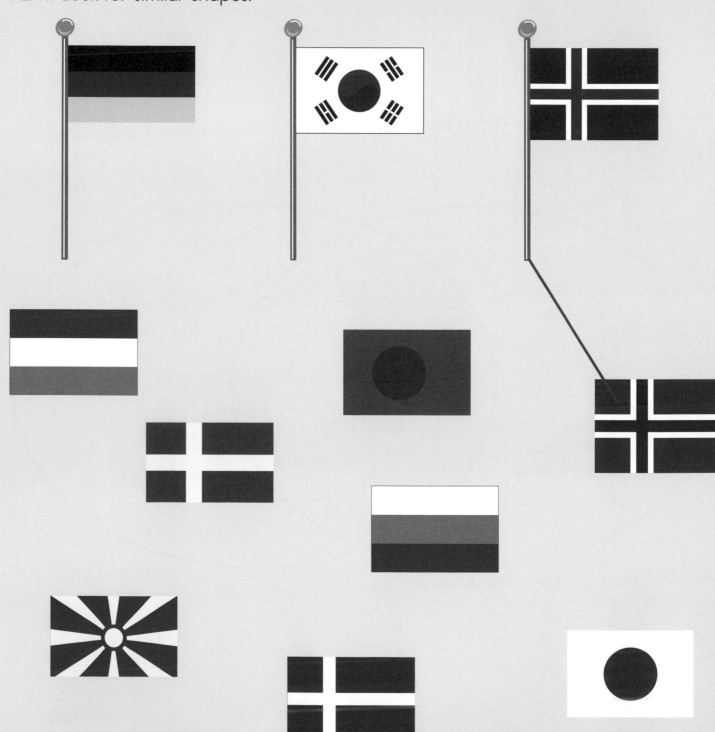

Squared Away

DRAW a line from each shape to its correct location.

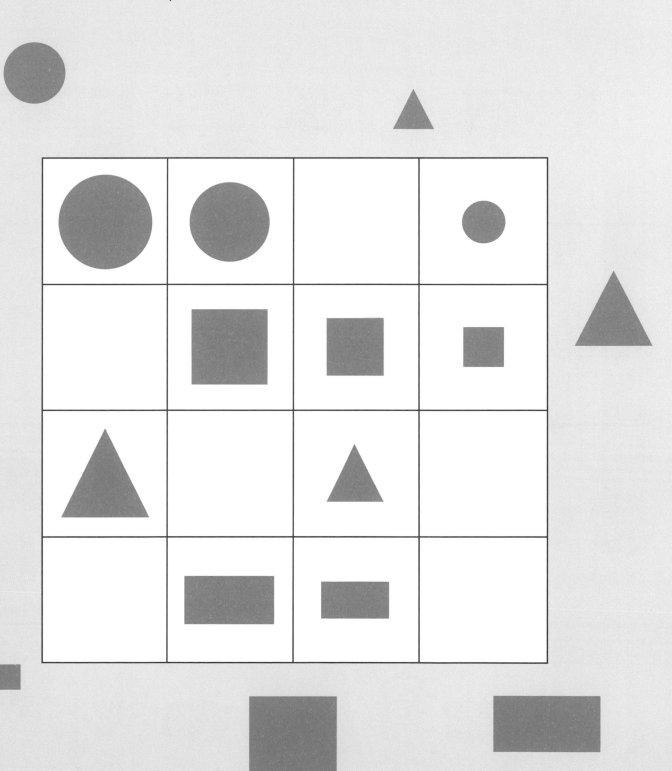

Review

What Comes Next?

CIRCLE the picture that comes next in the pattern.

1.

2.

3.

4.

Shapely Sequence

DRAW and COLOR shapes to finish each pattern.

Finish the Set

DRAW and COLOR a shape to finish each set.

HINT: There is more than one correct answer for each set. For an extra challenge, see how many shapes you can add to each set.

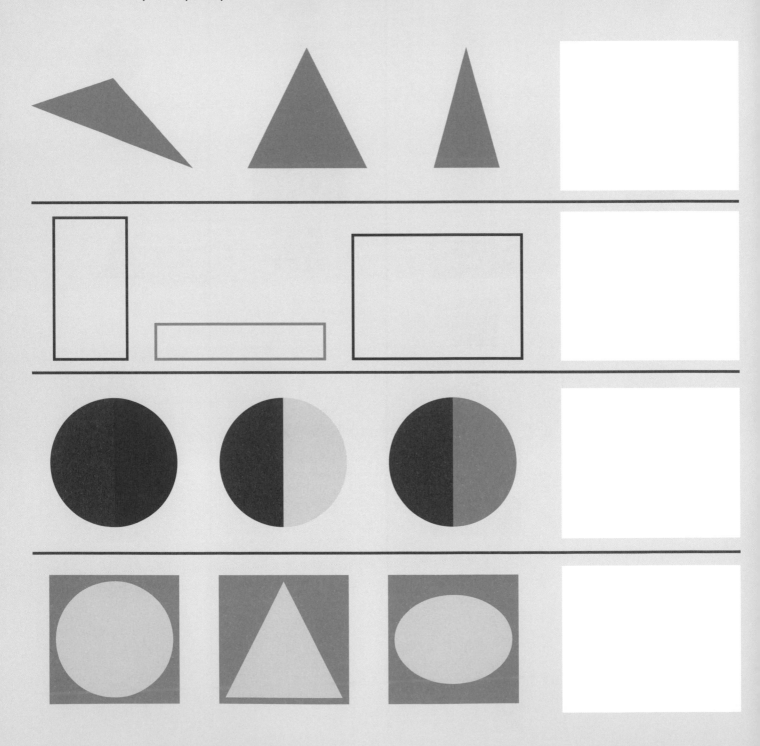

Squared Away

DRAW a line from each shape to its correct location.

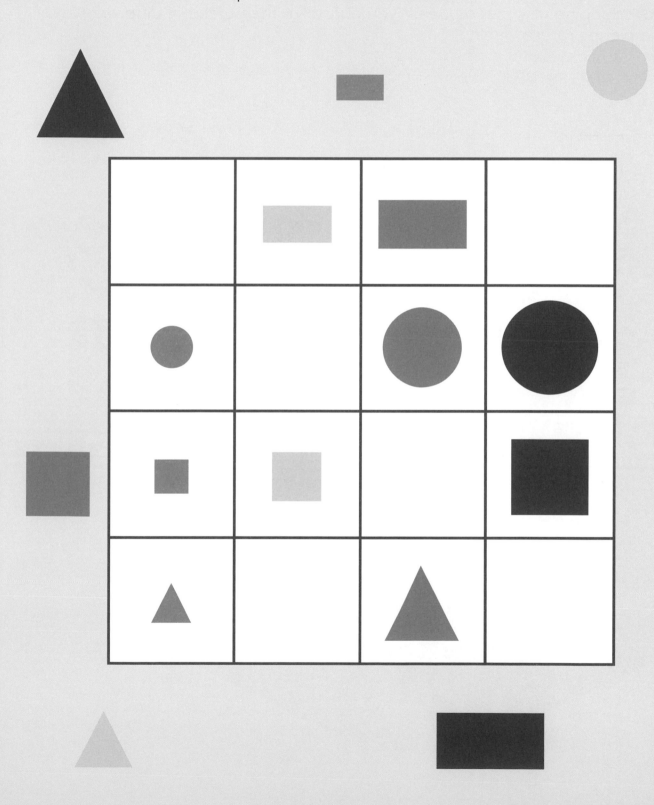

Shape Puzzles

Shape Puzzler

Using the shape pieces from page III, PLACE the pieces to completely fill each shape without overlapping any pieces. See if you can solve the puzzles in different ways. (Save the pieces to use again.)

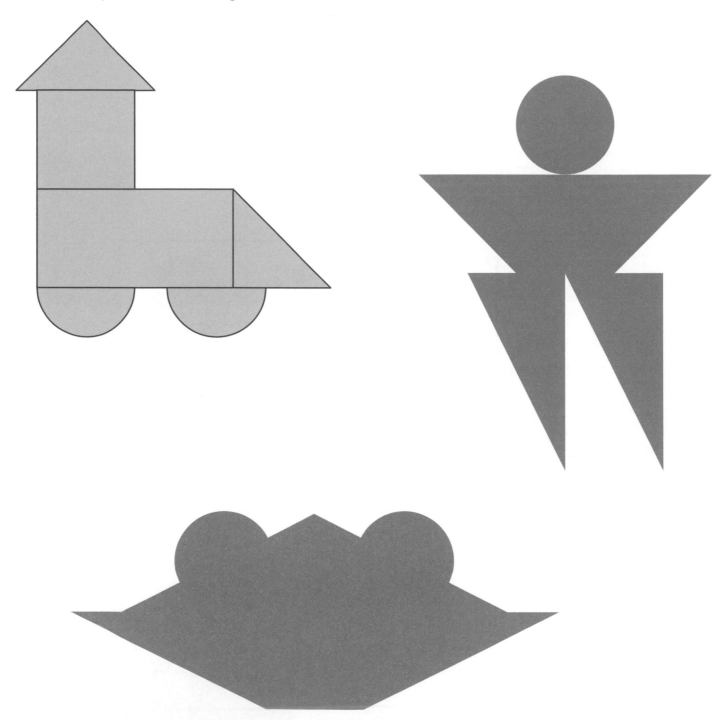

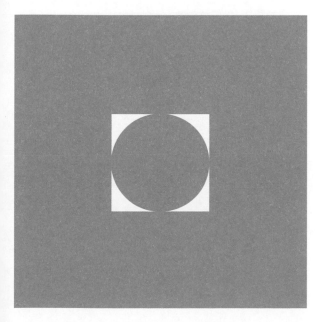

Shape Puzzles

Similar Shape

Using the shape pieces from page III, PLACE the pieces to fill each shape. Make a circle and square two different ways, and make a rectangle four different ways. (Save the pieces to use again.)

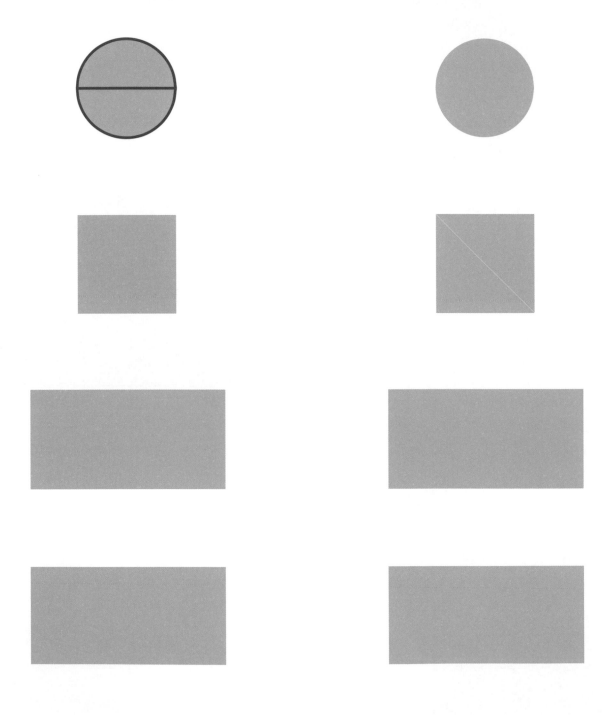

Match Up

DRAW a line to connect the shape pieces to the shape they will make when they are put together. Then PLACE the shape pieces from page III into the outlines to check your answers. (Save the pieces to use again.)

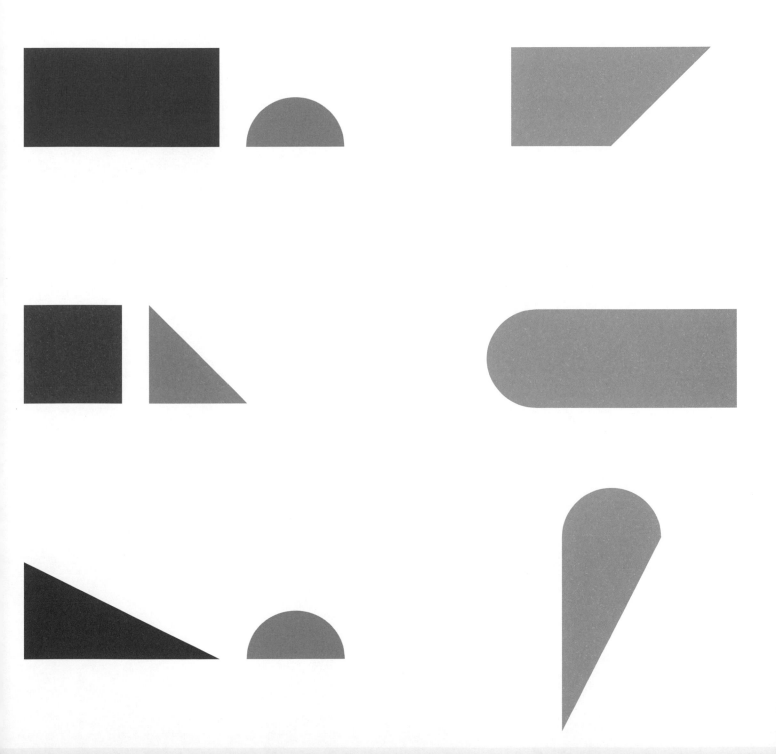

Shape Shifters

Using the pattern block pieces from page 113, PLACE the pieces to completely fill each shape without overlapping any pieces. See if you can solve the puzzles in different ways. (Save the pieces to use again.)

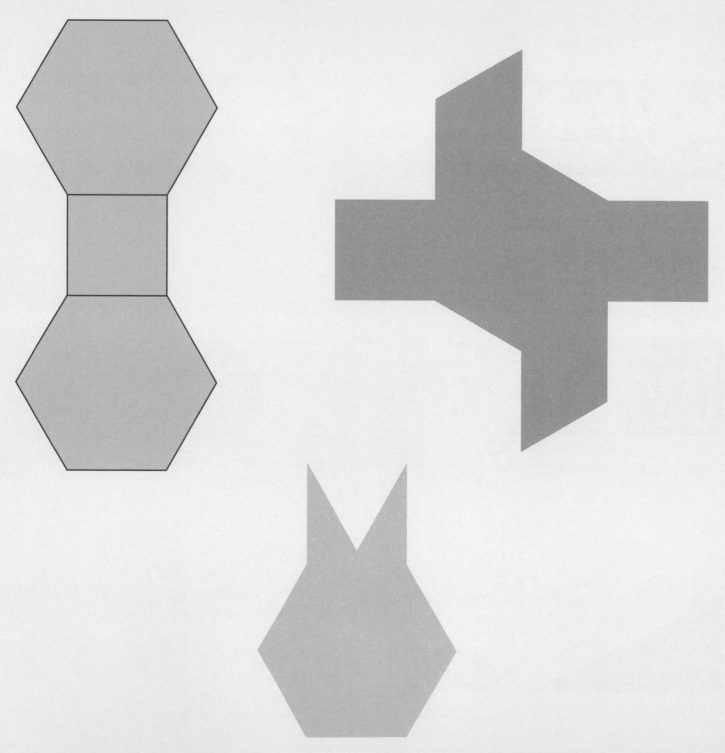

Shape Shifters

Using the pattern block pieces from page 113, PLACE the pieces to completely fill each shape without overlapping any pieces. See if you can solve the puzzles in different ways. (Save the pieces to use again.)

Tricky Tangrams

Using the tangram pieces from page 115, PLACE the pieces to completely fill each shape without overlapping any pieces. (Save the pieces to use again.)

HINT: Try placing the biggest pieces first. You do not need to use all of the tangram pieces in each shape.

Tricky Tangrams

Using the tangram pieces from page 115, PLACE the pieces to completely fill each shape without overlapping any pieces. (Save the pieces to use again.)

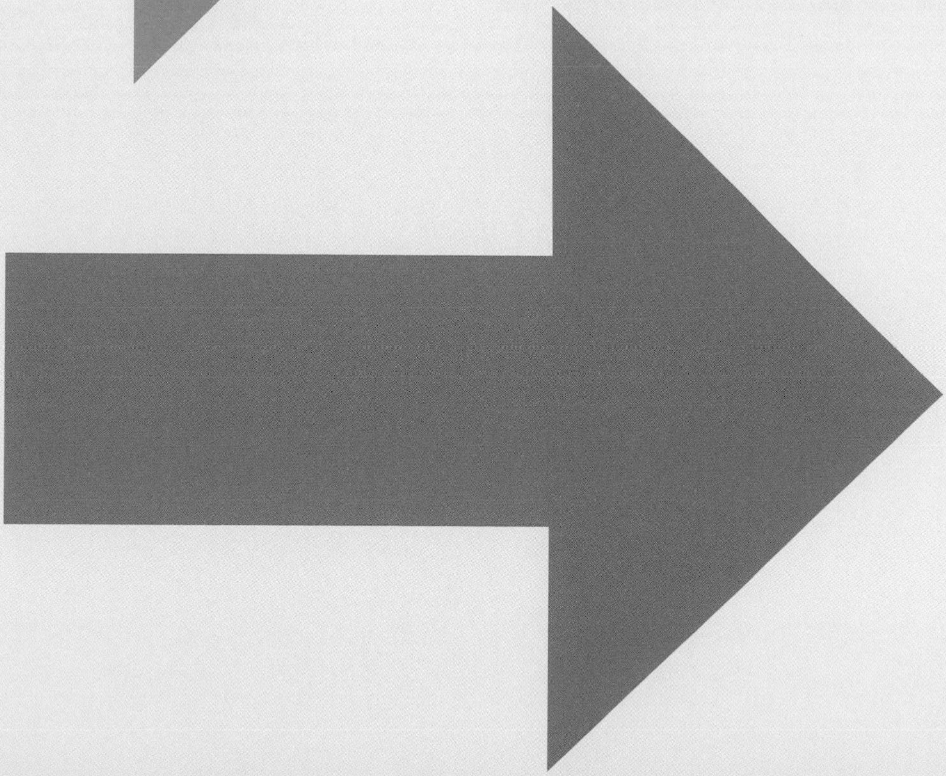

Shape Puzzler

Using the shape pieces from page 111, PLACE the pieces to completely fill each shape without overlapping any pieces. See if you can solve the puzzles in different ways. (Save the pieces to use again.)

Match Up

DRAW a line to connect the shape pieces to the shape they will make when they are put together. Then PLACE the shape pieces from page 111 into the outlines to check your answers.

Shape Shifters

Using the pattern block pieces from page 113, PLACE the pieces to completely fill each shape without overlapping any pieces. See if you can solve the puzzles in different ways.

Tricky Tangrams

Using the tangram pieces from page 115, PLACE the pieces to completely fill each shape without overlapping any pieces.

Location Words

Which One?

TRACE the circle that is **over** another shape.

Example: This circle is over the triangle.

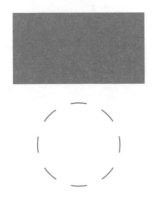

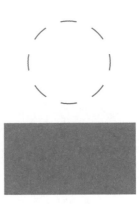

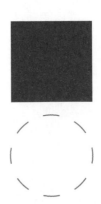

Doodle Pad

The fish is **under** the water. DRAW three more things under the water.

Flea Circus

FIND the flea in each picture. CIRCLE the pictures that show a flea **over** an object.

Store Shelves

FIND each box in the picture, and CIRCLE the cereal that is **under** it on the shelf.

1.

2.

3.

4.

Location Words

Which One?

TRACE the square that is **below** another shape.

Example: This square is below the triangle.

Doodle Pad

The cloud is **above** the house. DRAW three more things above the house.

Shapes Squared

FOLLOW the directions, and DRAW and COLOR the shapes in the correct squares on page 103.

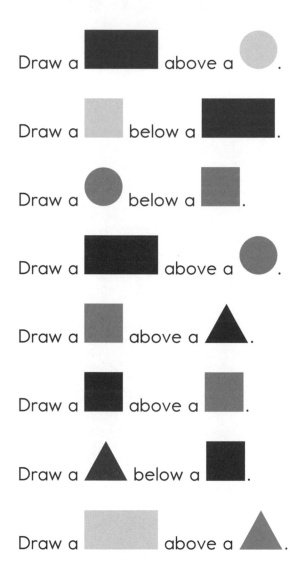

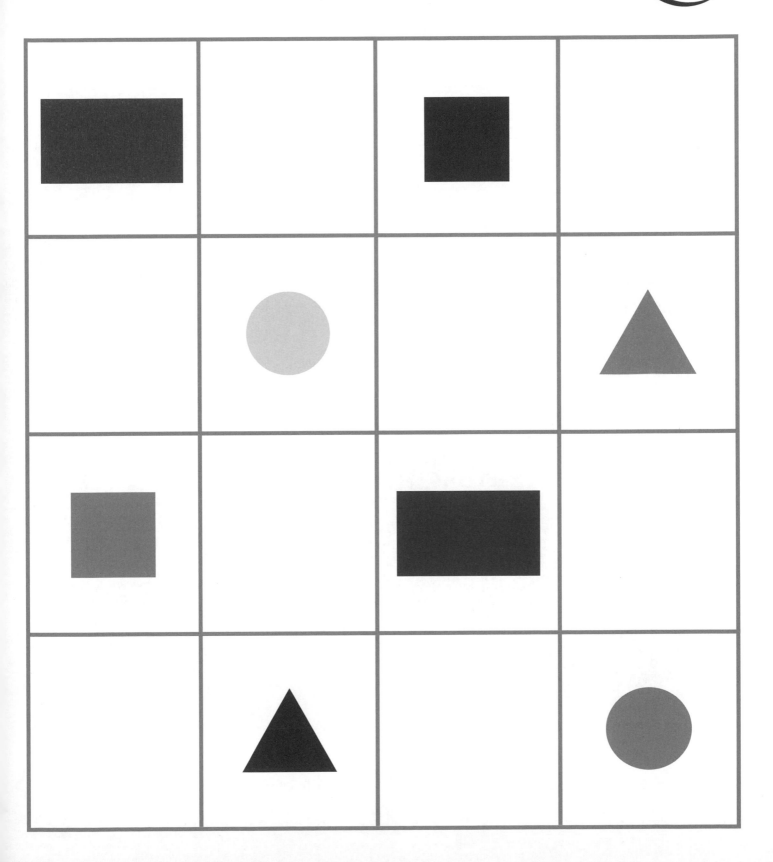

Park Party

There is a party in the park at one of the three picnic areas. FOLLOW the directions, and CIRCLE the correct party location.

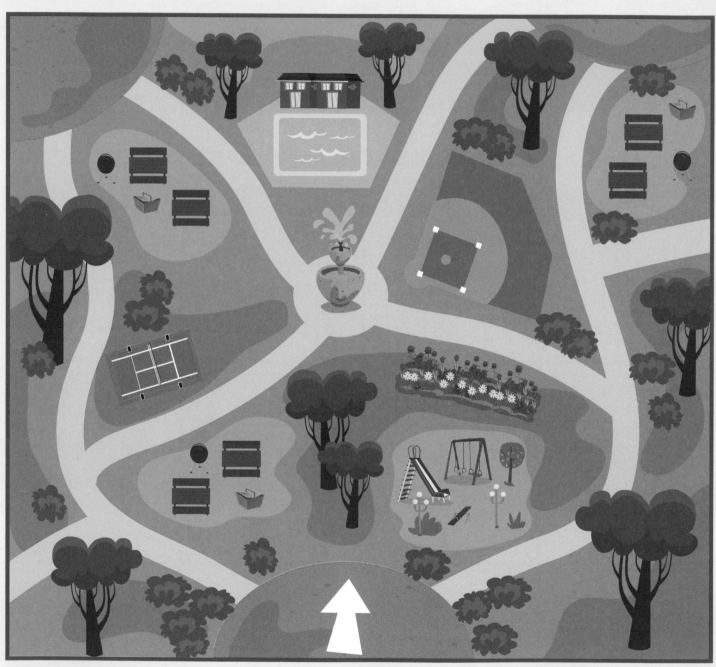

When you enter the park, go toward the 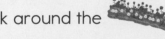. Walk around the toward the .

Take the path that is not by a or a , and you will find the party at the .

Treasure Hunt

There's only one pirate treasure. Some are fakes! FOLLOW the pirate's directions, and DRAW an X on the correct treasure.

Walk from my through the . Walk around the and cross the 2nd you find. By the you will find a . There you will find my .

Where Am I?

FOLLOW the path of each person on the map. Then CIRCLE the final location of that person.

1.

I left my and rode around the  . I crossed the first street to

get where I was going. Where am I?

2.

We left our and turned by the  to stop at the .

Then we rode around the corner and crossed the street after seeing

the to get where we were going. Where are we?

3.

I left my and stopped to get gas at the .

Then I went back past my and turned away from my house by

the . I did not make any more turns. Where am I?

Shapes Squared

FOLLOW the directions, and DRAW and COLOR the shapes in the correct squares.

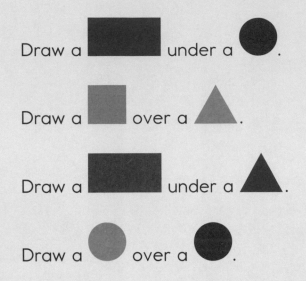

Draw a ▬ under a ●.

Draw a ◼ over a ▲.

Draw a ▬ under a ▲.

Draw a ● over a ●.

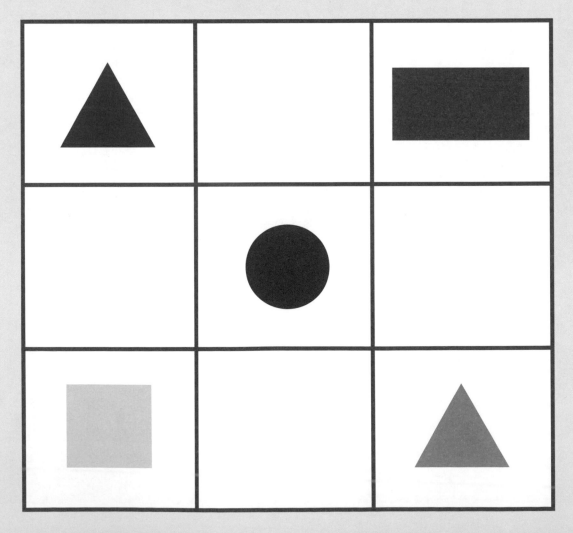

Which One?

COLOR red the triangle that is **above** another shape. COLOR blue the triangle that is **below** another shape.

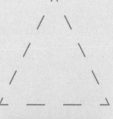

Where Am I?

FOLLOW the path of the boy on the map. Then CIRCLE his final location.

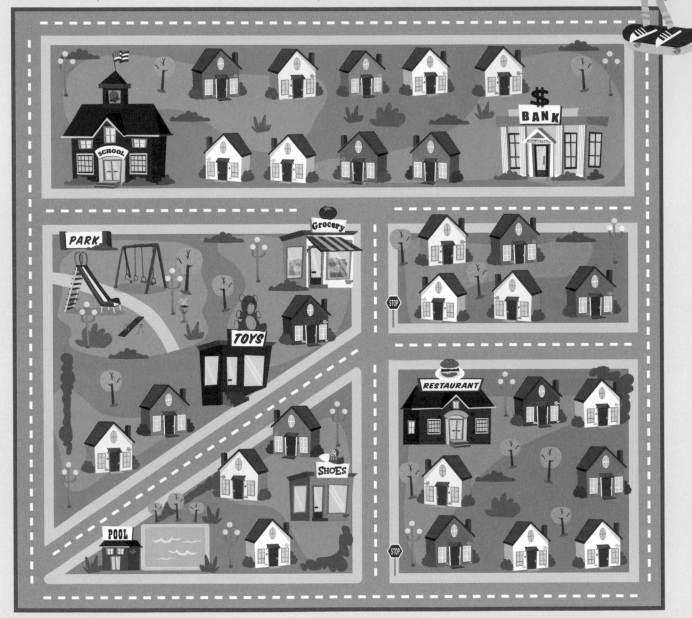

I left my  and turned to go around the . I turned when I got to the , heading toward the . I ended up on the block shaped like a triangle.

Where am I?

Shape Pieces

CUT OUT the 42 shape pieces.
These shape pieces are for use with pages 80, 81, 82, 83, 92, and 93.

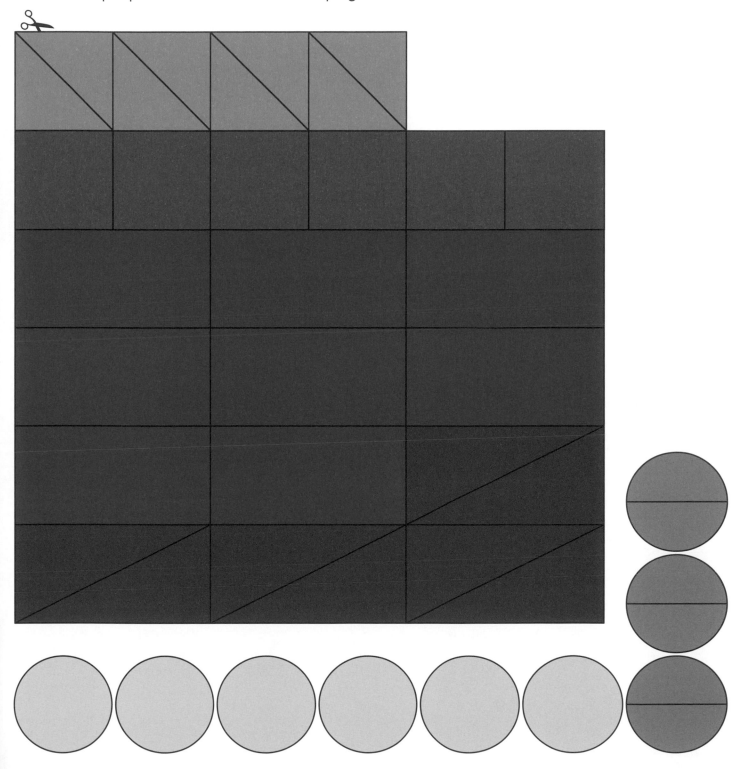

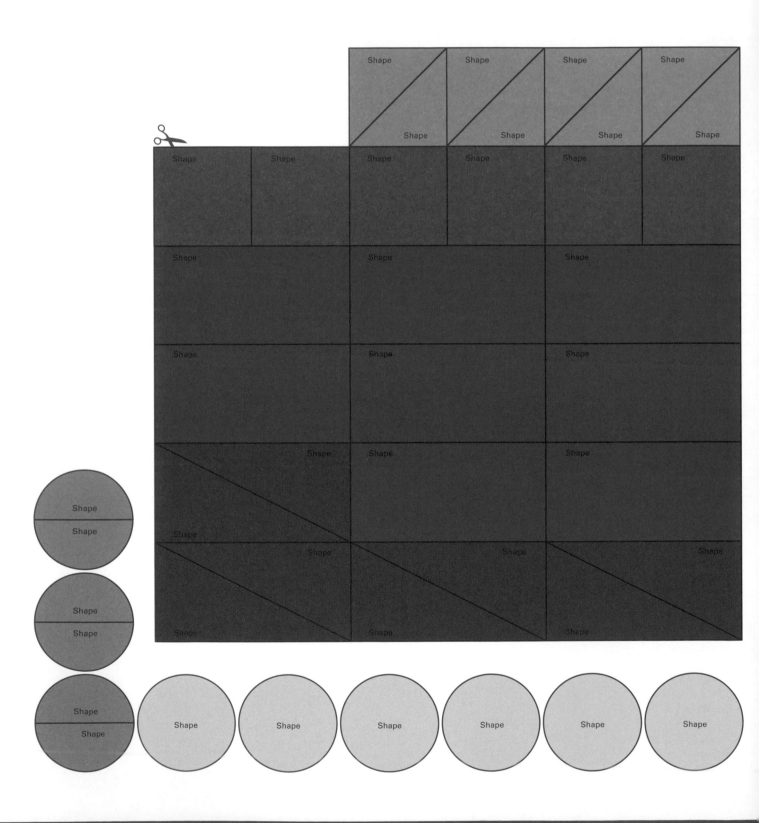

Pattern Blocks

CUT OUT the 31 pattern block pieces.
These pattern block pieces are for use with pages 84, 85, 86, 87, and 94.

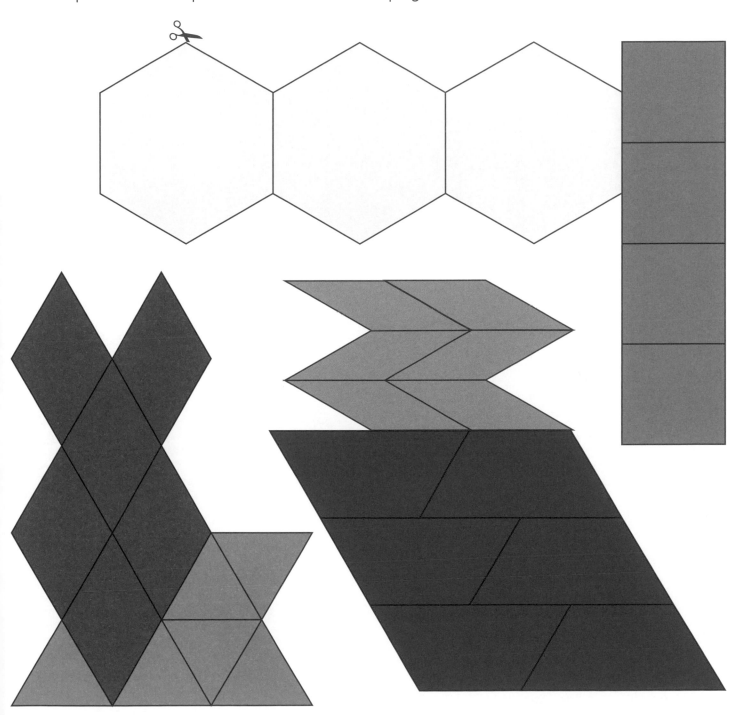

Game Pieces

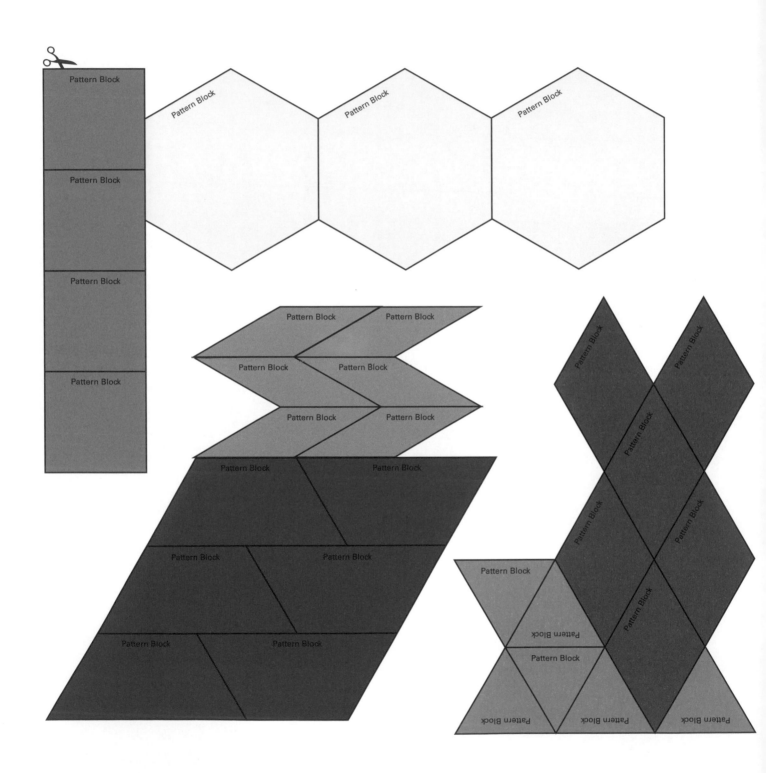

Pattern Block

Pattern Block

Pattern Block

Pattern Block

Pattern Block

Pattern Block

Pattern Block

Pattern Block

Pattern Block

Pattern Block

Pattern Block

Pattern Block

Pattern Block

Pattern Block

Pattern Block

Pattern Block

Pattern Block

Pattern Block

Pattern Block

Pattern Block

Pattern Block

Pattern Block

Pattern Block

Pattern Block

Pattern Block

Pattern Block

Pattern Block

Pattern Block

Pattern Block

Tangrams

CUT OUT the seven tangram pieces.
These tangram pieces are for use with pages 88, 89, 90, 91, and 95.

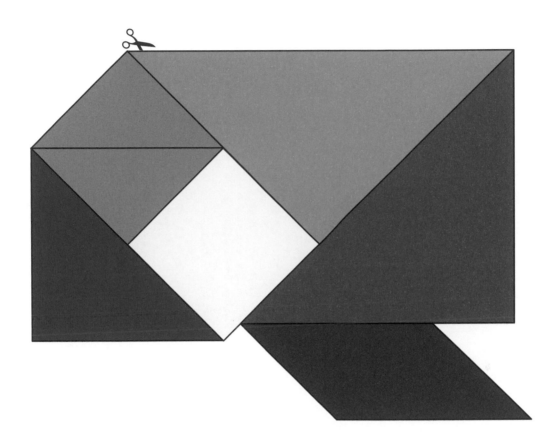

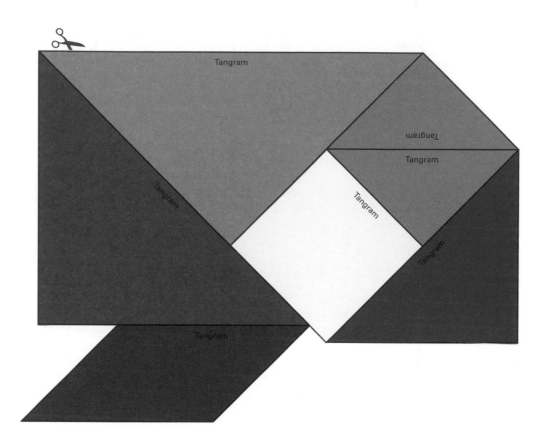

Answers

Page 3

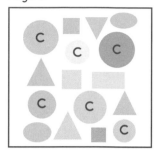

Page 9

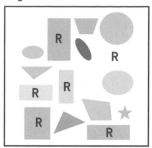

Page 15

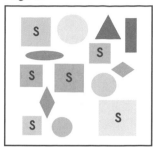

Page 21

Page 4

Page 10

Page 16

Page 22

Page 5

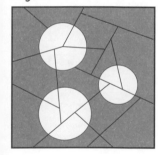

Page 11

Page 17

Page 23

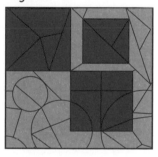

Page 6

Page 12

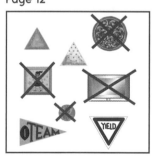

Page 18

Page 24

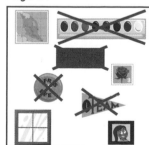

Page 7
Have someone check
your answers.

Page 13
Have someone check
your answers.

Page 19
Have someone check
your answers.

Page 25
Have someone check
your answers.

Answers

Page 26

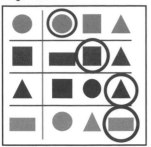

Page 27

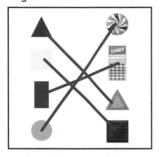

Page 28

Page 29

1. 5
2. 4
3. 2

Page 30

Page 31

Page 32

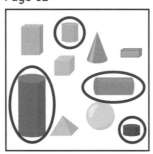

Page 33

Page 34

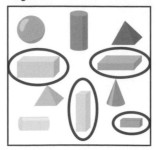

Page 35

Page 36

Page 37

Page 38

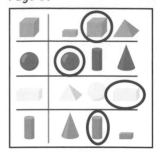

Page 39

Page 40

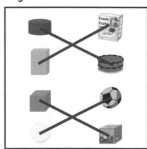

Page 41

1. 6
2. 4
3. 3

Answers

Page 42

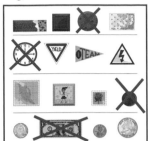

Page 43

Have someone check
your answers.

Page 45

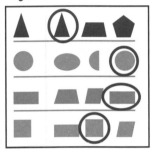

Page 46

Page 47

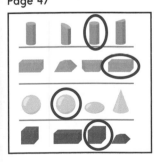

Page 48

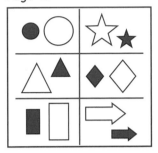

Page 49

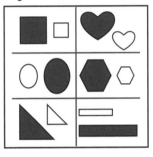

Page 50

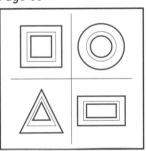

Page 51

Page 52

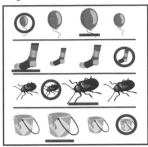

Page 53

Page 54

Page 55

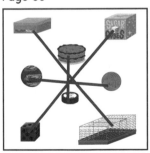

Page 56

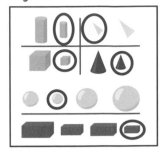

Page 57

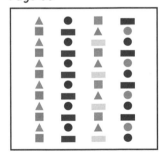

Pages 58–59

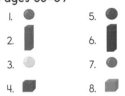

1.
2.
3.
4.
5.
6.
7.
8.

Page 60

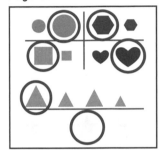

Page 61

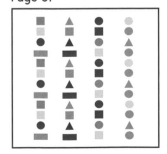

Answers

Page 62

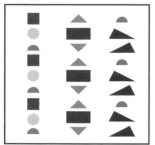

Page 63

Page 64

Suggestion:

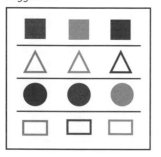

Page 67

Suggestion:

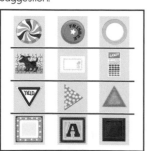

Page 68

Page 69

Suggestion:

Page 70

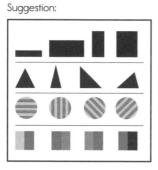

Page 71

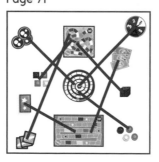

Page 72

Page 73

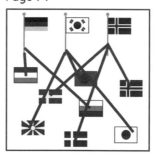

Page 74

Page 75

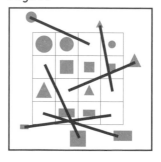

Page 76

1.   2.

3.   4.

Page 77

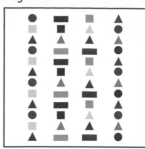

Page 78

Suggestion:

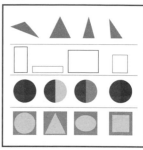

Page 79

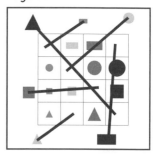

Page 80

Suggestion:

Page 81
Suggestion:

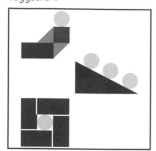

Page 82
Suggestion:

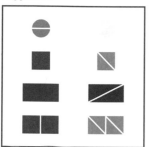

Page 83

Page 84
Suggestion:

Page 85
Suggestion:

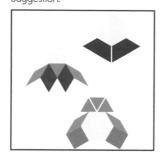

Page 86
Suggestion:

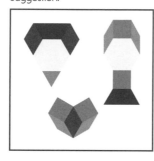

Page 87
Suggestion:

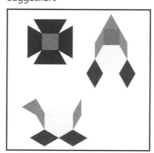

Page 88
Suggestion:

Page 89
Suggestion:

Page 90
Suggestion:

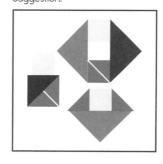

Page 91
Suggestion:

Page 92
Suggestion:

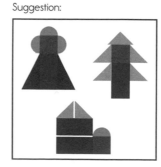

Page 93

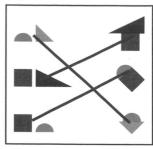

Page 94
Suggestion:

Page 95
Suggestion:

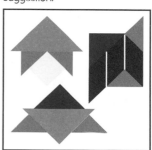

Page 96

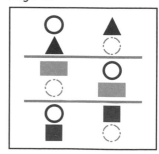

Page 97
Have someone check
your answers.

Answers

Page 98

Page 99

Page 100
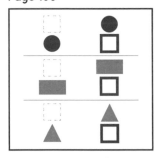

Page 101
Have someone check
your answers.

Pages 102–103
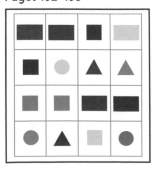

Page 104

Page 105
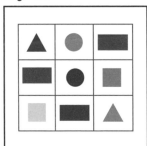

Pages 106–107
1.
2.
3.

Page 108
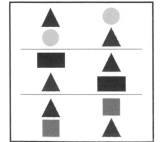

Page 109

Page 110